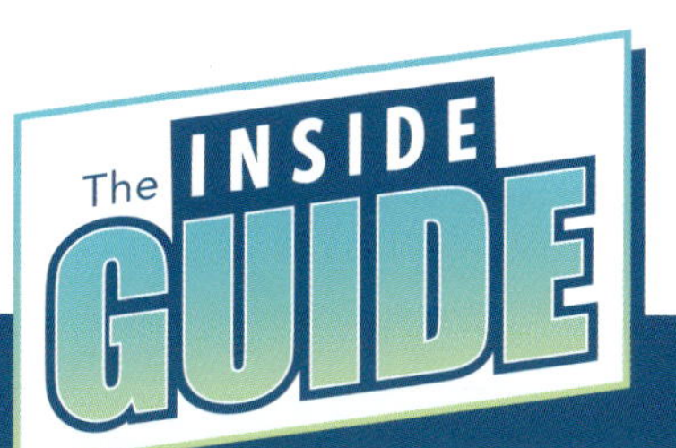

BIOLOGY BASICS

Life Cycles

By Leigh McClure

Published in 2025 by Cavendish Square Publishing, LLC
2544 Clinton Street Buffalo, NY 14224

First Edition

Cataloging-in-Publication Data

Names: McClure, Leigh.
Title: Life cycles / Leigh McClure.
Description: Buffalo, NY : Cavendish Square Publishing, 2025. | Series: The inside guide: biology basics | Includes glossary and index.
Identifiers: ISBN 9781502673237 (pbk.) | ISBN 9781502673244 (library bound) | ISBN 9781502673251 (ebook)
Subjects: LCSH: Life cycles (Biology)–Juvenile literature.
Classification: LCC QH501.M28 2025 | DDC 571.8–dc23

Editor: Caitie McAneney
Copyeditor: Nicole Horning
Designer: Deanna Lepovich

The photographs in this book are used by permission and through the courtesy of: Cover Darkdiamond67/Shutterstock.com; p. 4 Mark Brandon/Shutterstock.com; p. 6 MattL_Images/Shutterstock.com; p. 7 895Studio/Shutterstock.com; pp. 8, 27 fizkes/Shutterstock.com; p. 9 FOTOGRIN/Shutterstock.com; p. 10 PeskyMonkey/Shutterstock.com; pp. 12, 26 BlueRingMedia/Shutterstock.com; p. 13 (main) anmbph/Shutterstock.com; p. 13 (inset) New Africa/Shutterstock.com; p. 14 (main) Heather Lucia Snow/Shutterstock.com; p. 14 (inset) Alexapicso/Shutterstock.com; p. 15 Stephen Moehle/Shutterstock.com; p. 16 Mike Truchon/Shutterstock.com; p. 17 (left) Tyler Fox/Shutterstock.com; p. 17 (right) Carlos Pereira M/Shutterstock.com; p. 18 Drp8/Shutterstock.com; p. 19 Kathy Clark/Shutterstock.com; p. 20 Galyna Andrushko/Shuttertstock.com; p. 21 (top) Dan_Koleska/Shutterstock.com; p. 21 (bottom) Vishnevskiy Vasily/Shutterstock.com; p. 22 Plouy Thierry/Shutterstock.com; p. 23 Sean Barlow/Shutterstock.com; p. 24 slowmotiongli/Shutterstock.com; p. 25 Inara Prusakova/Shutterstock.com; p. 28 (top) metamorworks/Shutterstock.com; p. 28 (bottom) by-studio/Shutterstock.com; p. 29 (top) Rui Manuel Teles Gomes/Shutterstock.com; p. 29 (bottom) Suphalak Rueksanthitiwong/Shutterstock.com.

Some of the images in this book illustrate individuals who are models. The depictions do not imply actual situations or events.

CPSIA compliance information: Batch #CWCSQ25: For further information contact Cavendish Square Publishing LLC at 1-877-980-4450.

Printed in the United States of America

CONTENTS

A moth's complete change from a caterpillar into a winged adult is called metamorphosis.

Chapter One

THE CIRCLE OF LIFE

How can you tell if something is a living thing? To be alive, it must have the ability to grow and change, and eventually die. This is true of plants, fungi, bacteria, animals, and of course, humans.

The series of changes that happens to a living thing as it grows up is called a life cycle. A life cycle is predictable, because it happens to every member of a **species**. For example, a bird starts life as an egg, then breaks out of its shell, grows feathers, and eventually flies on its own. Nearly all birds go through these steps, or stages.

As things age, their features often change. Some species make **extreme** changes, such as that of a caterpillar into a moth or butterfly.

All About Cells

Each living thing's life begins with a cell. Cells are the basic units of all living things. Small organisms, or living things, like bacteria have only one cell. Others have many cells, sometimes billions or trillions. Most are specialized, or have certain jobs. For example, red blood cells deliver oxygen to body parts.

Fast Fact

Plant cells have a cell wall. Animal cells do not. This gives them different shapes.

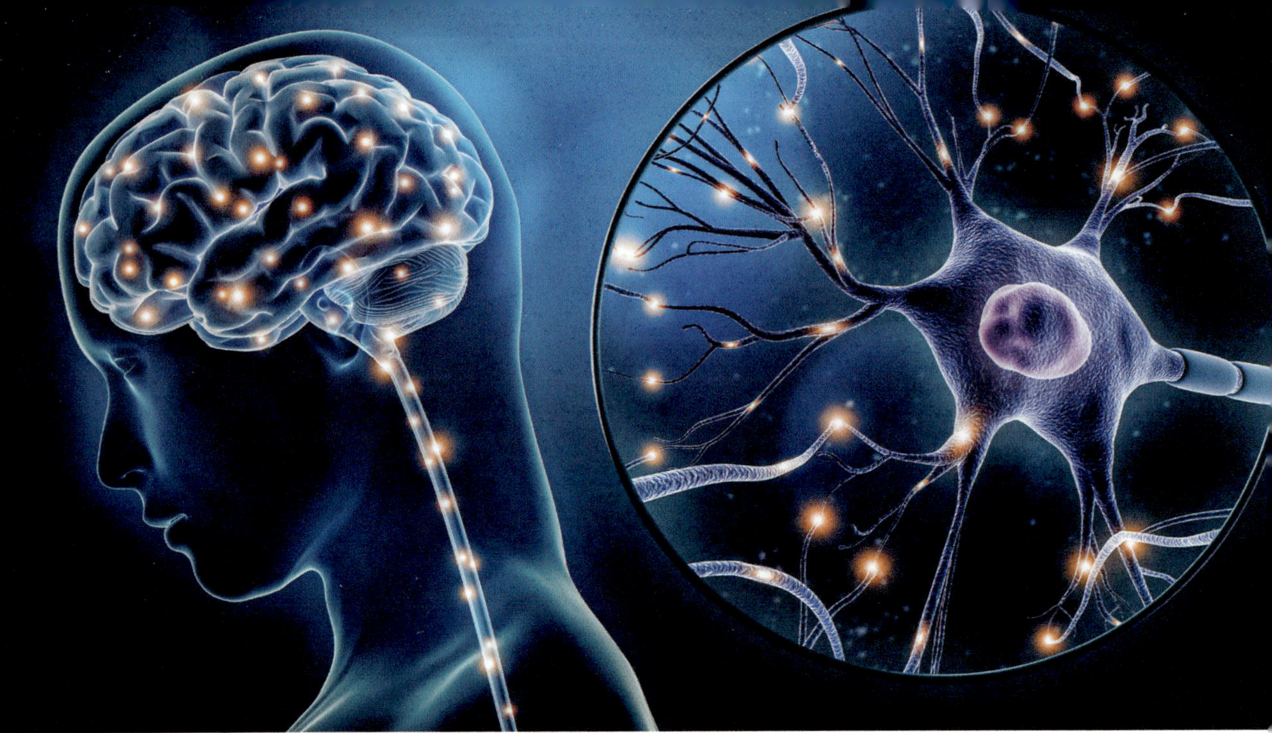

Neurons are cells that send messages from your brain to the rest of your body.

Cells create the proteins that carry out work in the body. They determine how a living thing moves, grows, and survives in its environment. Just like all living things, cells are created, grow, and die. Think of dry skin on your body; those flakes contain dead skin cells. Proper nutrition and enough water help cells live longer, which can help an organism thrive.

A Body Blueprint

Within a cell, there is a special part called a nucleus, which acts like the control center of the cell. There are chromosomes there, which contain DNA, or deoxyribonucleic acid. DNA is like a blueprint of a living thing. It makes up an organism's genes, which control what it will look like, how it will develop, and how long it may live naturally. If a life cycle was a movie, this would be a loose script.

REPRODUCTION

When two organisms of the same species create offspring, that's called sexual reproduction. The offspring receives half of its genetic information from one parent and half from the other parent. The resulting cell with this new genetic information divides into a copy of itself. Then, it divides again. It keeps dividing, becoming an **embryo**. The cells form specialized parts, grow into a body, and eventually, the living thing is born. The new organism may look and act similar to its parents, but it has its own DNA. When it's developed enough, it may also reproduce, passing its genes on to offspring.

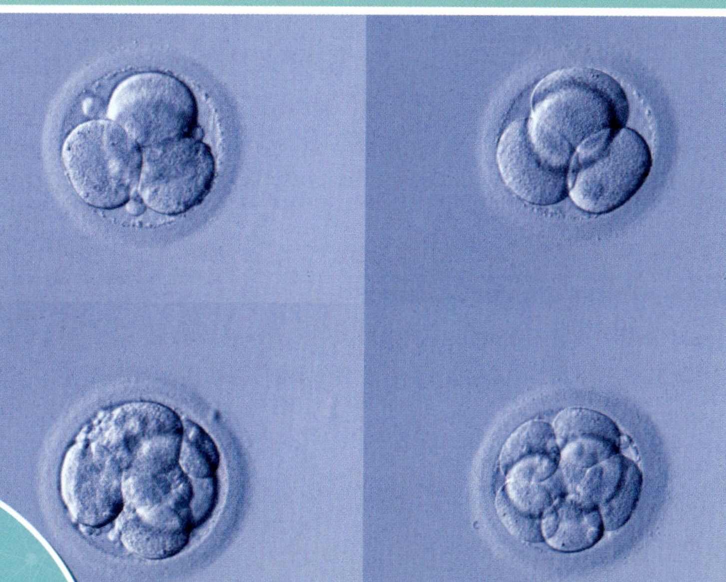

When a sperm cell meets an egg cell, it makes a **fertilized** egg. That cell divides quickly into an embryo.

Fast Fact

Some living things can reproduce all on their own. This is true for bacteria and some other organisms, such as certain kinds of lizards.

Life cycles are true across a species, from generation to generation.

The organisms in a species have slightly different DNA from one another. However, their lives all generally follow the same cycle. Think of your own life. You may not look exactly like your parent, but you both started life as a baby and grew into a child. Someday, you will be an adult, just like those who went before you.

How Long Is a Life?

While all living things are born, grow, and die, the length of their life cycles are different. The code for their possible lifespan is in their DNA.

The animal with the shortest lifespan is a mayfly. They live only around one day!

People live much longer than mayflies. Human **life expectancy** around the world is around 73 years old. This lifespan is much shorter than some animals, though. Giant tortoises can live for close to 200 years. Some sea sponges live for more than 10,000 years!

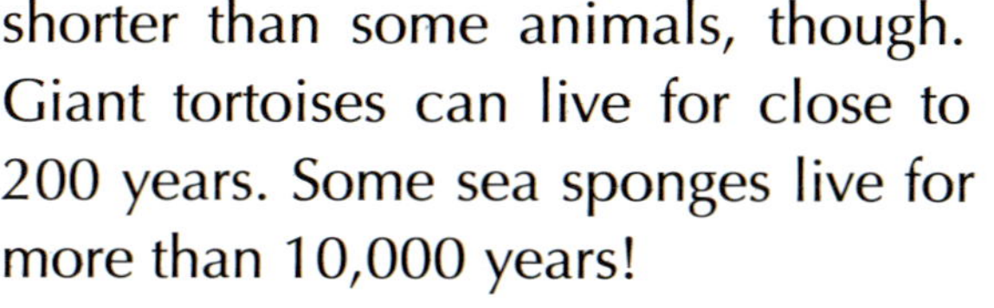

Fast Fact

Sometimes males and females of a species have different lifespans. Queen ants can live up to 15 years, but male ants live only a week or two.

Giant tortoises are known for their long lifespans.

After reproduction occurs, a seed forms. The largest seed on Earth is the coco de mer.

PLANT LIFE CYCLES

Have you ever watched a plant sprout through the soil? Maybe you have houseplants around your home that you care for. These are great opportunities to observe plant life cycles. Plants follow different paths depending on their species, but they all start with **spores** or seeds.

Plants use photosynthesis to give them energy to grow. In photosynthesis, plants use carbon dioxide and sunlight to produce oxygen and a sugar called glucose. With the right amount of water, carbon dioxide, sunlight, as well as the right kind of soil, plants will grow. Eventually, their cells will die, and they will return to the earth.

Flowering Plants

In flowering plants, flowers allow the plants to reproduce. Pollen from one plant's male parts is brought to another plant's female parts. Then, a fruit starts to grow, and within the fruit, there are seeds. Seeds are the embryos of plants, with a hard shell for protection.

By wind, rain, animals, or other forces, seeds end up in the soil. A seed starts to grow

Fast Fact

Plant spores are reproductive units made of a single cell that begin the life cycle of plants such as mosses and ferns.

PLANT PARTS

Plants have roots that take in water from the soil and stems for support. They also have leaves that act as the food-making parts. Many plants have bright, sweet-smelling flower petals. These attract insects, which increases the plant's chances of pollination. The male part of a flowering plant is a stamen, which has an anther to make pollen. The female part is the pistil, which includes an ovary. When pollen touches the top of the pistil, it is sent down into the ovary. The joining of these male and female cells produces seeds.

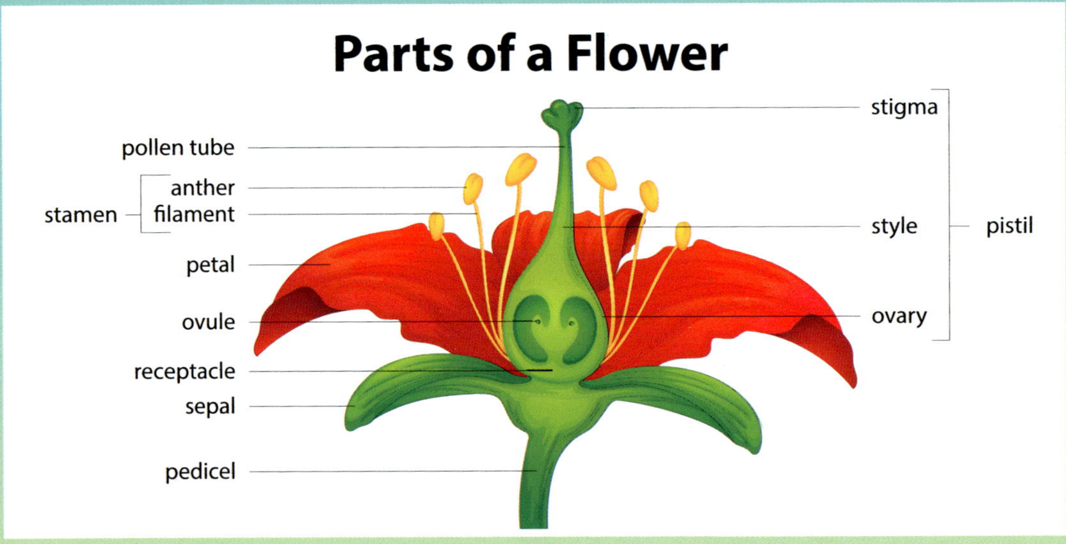

This flowering plant diagram shows the different parts that are important to reproduction.

roots down into the soil. Then, it starts to grow a stem, eventually pushing up through the soil as a sprout. With enough sunlight and water, the sprout grows into a mature plant. Mature plants have roots, stems, and leaves. The plant grows flowers, which make seeds, and the life cycle starts over.

Seasonal Trees

acorn

Many deciduous trees have broad, flat leaves. These include maple trees, oak trees, and elm trees. They drop seeds such as acorns, which sprout in the ground. They push up through the soil, first with a soft stem, and later, with a woody stem. They continue to grow each year. Their roots extend deeper into the earth to get more nutrients and water. Their branches reach farther outward as new cells grow.

Deciduous trees go through seasonal cycles as well as a life cycle. They shed their leaves each fall as the air gets

Fast Fact

Deciduous trees are found in temperate areas of Earth that have four seasons: winter, spring, summer, and fall.

The outer layer of a tree grows each year in the summer, pausing in the winter. This forms rings that can be counted to determine the tree's age.

colder. In the springtime, buds form on the branches. They open to reveal new leaves and shoots.

Coniferous Trees

Coniferous trees have hard, thin needles for leaves, and their seeds are found inside cones. Some coniferous trees—such

pine cone

Fast Fact

The longest-living tree in the world is called Methuselah, and it's more than 4,600 years old. It's a coniferous species called a Great Basin bristlecone pine.

Scientists looked at samples from Methuselah's core to guess its age.

Hyperion is more than 380 feet (115.8 meters) tall! It is off limits because visitors were beginning to ruin its surroundings.

as firs, pines, and blue spruces—keep their needles all year round. They're called evergreen trees.

Some coniferous trees have a superlong lifespan and exceptional heights. The tallest living tree in the world is named Hyperion, a redwood tree found in California's Redwood National Park.

Unlike plants, animals cannot make their food through photosynthesis so they need to grow and change throughout their lives to get the food they need.

ANIMAL LIFE CYCLES

Animals and plants are both living things, but they are also very different from one another. Unlike plants, animals usually have **mobility**, which allows them to find food and mates for reproduction. The animal kingdom is broken down into several groups, which include insects, reptiles, amphibians, birds, fish, and mammals.

Insect Life Cycles

Insects start their lives inside an egg. Their parent usually lays many eggs at once. The insect eventually breaks out of the egg. In some cases, it is a nymph, which looks like a smaller version of an adult, while in other cases, it is a larva that looks a bit like a worm.

As nymphs grow, they must cast off their **exoskeleton**. This is called molting. They may grow wings. For larvae of insects like butterflies and

There are over a million different species of insects in the world! Grasshoppers start their lives as nymphs, while butterflies start as caterpillars.

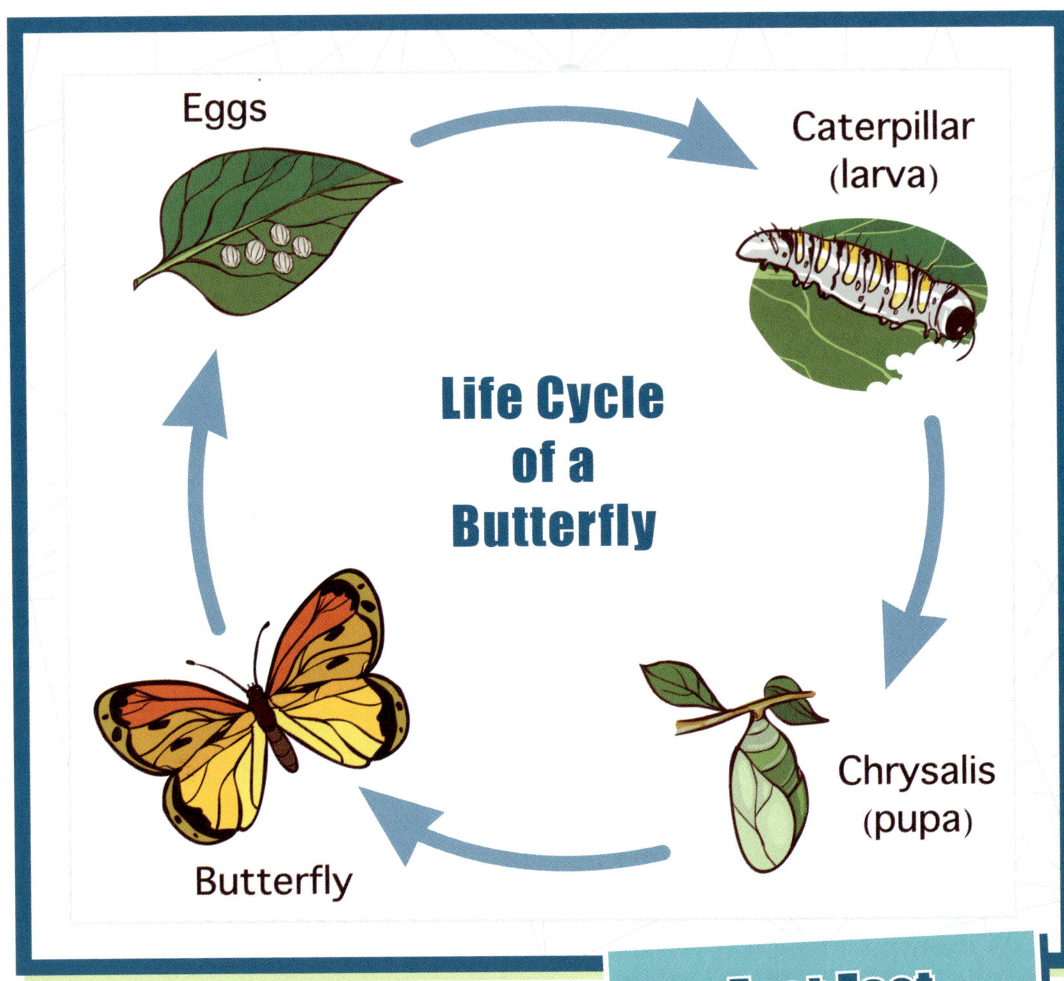

As butterflies go through their life cycle, their bodies change completely. They grow broad wings that later help them fly from flower to flower for food.

moths, they turn from a larva into a pupa. They hide inside a cocoon where their bodies become like a soup and reorganize into their final adult form.

Fast Fact

Insects are arthropods, which are creatures characterized by hard exoskeletons. The arthropod family also includes spiders and crustaceans (like crabs).

CICADAS

Cicadas have a very unusual life cycle. That's because some species, called periodical cicadas, spend much more of their lives underground as nymphs than aboveground as adults. Female cicadas lay their eggs in trees, where they hatch. Then, the nymphs fall from the tree and tunnel into the ground. Some nymphs stay in the ground for 17 years! When they finally come out, they live for only around four weeks. Their main job in this time is to mate. A group of cicadas coming out of the earth is called a brood. In 2024, two huge broods of 13- and 17-year cicadas were expected to come out together for the first time in over 220 years.

Male cicadas make very loud noises when they're looking for a mate. When they all "sing" together, it can be as loud as a jet plane.

Fish Life Cycles

More than 30,000 species of fish live in bodies of water around the world. A small percentage of fish give birth to live young, but most

Salmon take a special journey in their life cycle. They swim from fresh water to the ocean, then return months or years later to spawn.

fish start their lives as eggs. Female fish release many eggs into the water, which is called spawning. Male fish release a fluid called milt that fertilizes the eggs. If the egg survives, a larval fish comes out of it. It gets its food from a yolk sac on its body. When it can eat on its own, it's called a fry. As a fish gets older, it grows into a juvenile fish. Eventually, it matures and develops the ability to reproduce. When a female lays eggs, a new life cycle begins.

Fast Fact

Some sharks—such as hammerheads and bull sharks—are viviparous, which means that they give birth to live young.

Reptiles and Amphibians

Most reptiles and amphibians are **cold-blooded** and begin their lives as eggs, but their life cycles are very different. Reptiles, such as snakes and lizards, lay eggs with a tough shell on land. When they break out of their shell, they already have lungs to breathe the air. Most look like small versions of adult reptiles.

Amphibians such as frogs and toads lay their soft eggs in water. When they hatch, amphibian larvae live in the water for a while.

They breathe underwater using gills and swim with tails. Lungs and legs grow as an amphibian develops. Eventually, it can live on land.

Many reptiles can survive in areas with very little water, while amphibians need water.

Fast Fact

Boas and vipers give birth to live young instead of laying eggs.

Bird Life Cycles

Birds are also born from hard-shelled eggs. A mother bird often lays her eggs in a nest, then keeps them warm with her body heat. When the birds hatch, they lack feathers and eyesight. The mother usually gets them food and keeps them safe until they can fly.

Once a bird grows its flight feathers, it's called a fledgling. It practices leaving the nest but can't fly too far yet. A juvenile bird is able to fly and get its own food. Adult birds can reproduce, laying their own eggs and starting a new life cycle.

Cuckoos have a strange start to their life cycle. Some cuckoo parents hide their eggs in the nests of other birds. The cuckoos are fed and raised by another species!

Wolves live in tight family groups called packs. Mothers feed their babies and keep them safe.

HUMANS AND OTHER MAMMALS

Mammals are animals that usually have fur or hair and give birth to live young. Earth is home to a great **diversity** of mammals. Some walk on land, while others swim. Some do a little of both. The African elephant can weigh up to 7 tons (6.35 metric tons), while tiny mice fit in the palm of your hand. Humans are part of the mammal family!

Unlike other kinds of animals, mammal mothers feed their babies milk. They care for them until they're ready to be out on their own. Many mammals grow up and live in family or community units, such as wolf packs, lion prides, and zebra herds. This is just one thing that humans have in common with other mammals.

echidna

Fast Fact

Mammals that reproduce by laying eggs are called monotremes. The only monotremes in the world are echidnas and the duck-billed platypus.

Mammal Life Cycles

Because mammals are so different from species to species, each life cycle looks a bit different. However, most mammals are born, stay with their mothers until they grow up, and then develop into adults with the ability to reproduce.

Some mammals are born in the water, such as whales and manatees. They can swim right away. Other mammals are born on land, such as lions and giraffes. Mothers often provide food for a while, then teach their young to gather or hunt food. Some mammals stay together when they grow up. Others, such as male tigers and elephants, may go off on their own or form looser connections when they become adults.

Fast Fact

Some rodents are born underground, such as naked mole rats and moles.

Mother lions teach their young how to hunt when they are just weeks old.

All About Babies

Much like other mammals, human babies grow inside their mothers. They start as collections of cells called embryos, then develop into fetuses. After around 40 weeks, a baby is born.

Unlike many animals, human babies are completely helpless at birth when it comes to movement. While baby horses and cows can walk soon after birth, human babies can't get around without their parents

Newborn babies sleep up to 19 hours a day!

Fetal Development

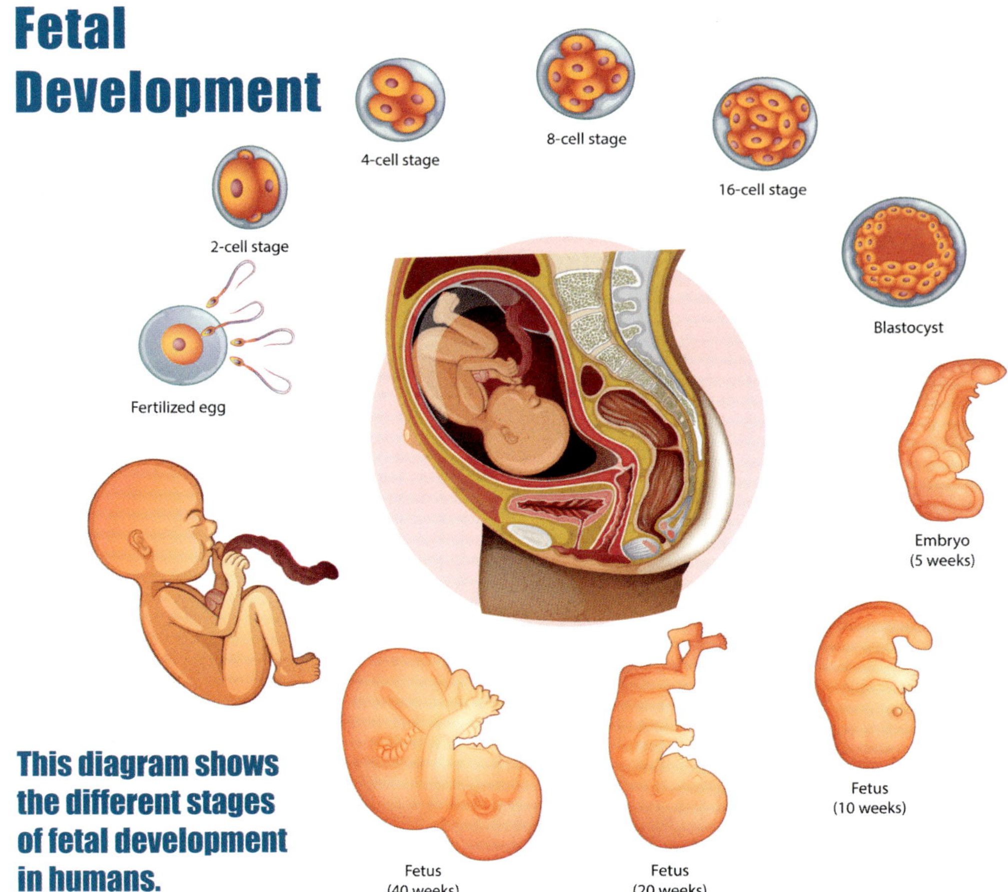

This diagram shows the different stages of fetal development in humans.

for many months. They rely on their caregiver for milk or baby formula.

Growing Up Human

For the first year, a human is considered an infant. Then, they are a toddler, able to get around on their own. At around age three, they are considered a child. They continue to grow and develop abilities that will help them later in life.

Fast Fact

Some babies are born preterm, or before 37 weeks of pregnancy. Preterm babies often spend time under close care in a hospital until they are developed enough to go home.

THE FINAL CHAPTER

The final step in every life cycle of every organism on Earth is death. When an animal or human is dying, the body starts to shut down. The vital organs—such as the brain and heart—stop working.

Sometimes this happens slowly because of disease or old age. Sometimes an organism dies quickly because of an injury. In nature, when something dies, it returns to the earth. Organisms called decomposers, such as earthworms and bacteria, feed on dead plants and animals. They take in nutrients from the dead organism and return them to the ecosystem. The circle of life continues.

The next life stage is puberty. This happens from ages 10 to 14 for females and 12 to 16 for males. Many physical changes happen to a person's body in this time. People with male sex characteristics develop deeper voices and facial hair. People with female sex characteristics start to **menstruate**. In puberty, a person develops the ability to reproduce. After this, they are seen an adult.

Puberty is a time of intense changes in a person's life. Having someone to talk to can help ease this step in the life cycle.

THINK ABOUT IT!

1. How does an organism's DNA affect its life cycle?

2. How can a change in climate affect the life cycle of a plant?

3. Why might fish and insects lay so many eggs at once?

4. How is the human life cycle similar to and different from other mammals?

GLOSSARY

cold-blooded: Having a body temperature that depends on the environment and is not regulated internally.

diversity: The quality or state of having many different types or forms.

embryo: An animal in the early stages of growth.

exoskeleton: An external supporting body structure.

extreme: More than what is usual or expected.

fertilize: To have reproductive cells come together and produce offspring.

generation: A group of living things born and living during the same time.

life expectancy: The average lifespan of an individual.

menstruate: To cyclically discharge blood, mucus, and dead cells from the uterus, as by a female of child-bearing age who is not pregnant.

mobility: The ability to move.

species: A group of plants or animals that are all the same kind.

spore: A tiny reproductive body made up of one or more cells.

FIND OUT MORE

Books

Jaycox, Jaclyn. *Unusual Life Cycles of Mammals*. North Mankato, MN: Capstone Press, 2022.

McDougal, Anna. *The Strange Life Cycle of a Mayfly*. Buffalo, NY: Scientific American Educational Publishing, 2025.

Websites

Britannica: "How Do Cicadas Know When to Go Aboveground?"
www.britannica.com/video/222430/did-you-know-17-year-cicada
Check out this video showing the 17-year cicada emerging from the earth.

The Frog Life Cycle for Kids
www.natgeokids.com/uk/discover/science/nature/frog-life-cycle/
Explore the stages of a frog's life with *National Geographic Kids*.

The Life Cycle of Flowering Plants
www.natgeokids.com/uk/discover/science/nature/the-life-cycle-of-flowering-plants/
Follow the life cycle of flowering plants with *National Geographic Kids*.

Publisher's note to educators and parents: Our editors have carefully reviewed these websites to ensure that they are suitable for students. Many websites change frequently, however, and we cannot guarantee that a site's future contents will continue to meet our high standards of quality and educational value. Be advised that students should be closely supervised whenever they access the internet.

INDEX